NERD GUIDE™ FOR CODING C#

I0011798

This work is Nerd Certified through *Nerd Guide's* Nerd Certification™ process.

To learn more about this and our other Nerd Guide publications please visit *http://www.nerdguide.org*

NERD GUIDE™ FOR CODING C#

EVENT DRIVEN PROGRAMMING, FIRST EDITION

PUBLISHED BY
Nerd Guide, P.O. Box 15559, Rochester, NY 14615

ISBN: 978-0-9887176-4-0

Revision 2; *Copyright © 2012 by Daniel Diaz III*

The cover artwork is of special concern as it took deep thought and thousands of seconds to work out the right visual message. Are your Nerd senses tingling yet?

While every precaution has been taken in the preparation of this book, the publisher assumes no responsibility for errors or omissions, or for damages resulting from the use of the information contained herein.

Microsoft, Microsoft .NET, MSDN, Visual C#, Visual Studio, Windows are either registered trademarks or trademarks of Microsoft Corporation.

*"Believe in yourself because few will do it for you.
Nevertheless, I will be one of those few."*
- Daniel Diaz III

Table of Contents

Chapter 1

Who Should Read This Guide

This should be read by people who are official Nerds. These people don't need fluff and overstated examples to get the point. Nerds are smart–so let's get to the point already. As an existing Nerd with some existing skill level, you need a certain amount of knowledge to get started in coding. Hopefully, this book is the right-sized nugget to put you on the path of adding a Visual C# feather to your cap.

TIP: Idiomatic expressions are words formed together that have a meaning that is different from the individual word definitions. Some expressions in this book have been added to encourage the use of Internet searching to explore the full meaning of each statement.

You will read about Visual C# event-driven programming for the web and for applications development. This book will guide you to tools to utilize and the knowledge base to work from. The information presented is focused on the popular uses of the Visual C# coding language. The goal is to keep the learning process to the point by handing you the keys to unlock the knowledge requirements to code. These keys will enable you to take this virtual world wherever you want, once you know what to look for.

You will not find specifics about coding best practices or architectural patterns. This material is here to help get you started - that is it. Your path to coding greatness will be illuminated but the comprehensive details needed to take each step must be found by you. We will address some topics like event-driven programming that nobody talks about but everyone wants to know about. Topics like this frankly are not well addressed, possibly because no one really knows. Here you will find the authors assessment which is the result of a lot of research, experimentation and Nerd know-how.

Consider your Nerd status, if you have one at all. Do you match any of these Nerd classes?

- Real Nerds, the kind who think this book might actually be fun to read;
- Those trying to pass themselves off as a senior Nerd;
- Those trying to pass themselves off as Visual C# Nerds;
- Those pretending to know something about being a code Nerd;
- Nerds pretending to know something about .NET coding;
- Those pretending to be web coders;
- Those trying to impress other Nerds.

Real Nerds - you know who you are. Yes, you are the people who pick up a quantum physics text book for bedtime reading. You are loathed people. Seriously, textbooks are not "fun" to read. One of your chromosomes must be out of whack. Someone should write an app to check for that.

To those who want to be senior programmers and think this book will somehow get you there: Keep reading, you might learn something.

TIP: Reading this work does not guarantee that you will: a) learn something; b) achieve Nerd status; or c) increase your existing status.

To those Visual C# wannabe coders, you too might learn something (this claim is not guaranteed). Be warned: coding in Visual C# is not something you become good at overnight. If your coding skills are low, expect to pay the piper before you gain the respect of fellow Nerds. I am talking about *years* of time invested, folks. Dream, work hard, read a lot, talk with other Nerds, and experiment. You will reach your goal, before you know it.

To those pretending to be coders, here is a secret: enterprise coding with Visual C# is *hard work*. It is not for the feeble. This Microsoft .NET stuff makes for a great buzz word, and additional skill for your Nerd *résumé*, but seriously think about starting some place easier. Learn to identify the areas you need to improve your knowledge development to get you from zero to hero status.

TIP: If you plan to remain a pretend coder you should really pretend to code on a different platform, something simpler. If you do not heed these words, a couple of novice interview questions might get you found out. Yes, some of them are in the book just for you.

To all those Microsoft .NET wannabes–this material is here to help. Do not be put off by the lashings offered to these other Nerd groups. Follow the learning patterns and you may find your way.

TIP: If you cannot conceptualize what is meant by "learning pattern" then just stop reading now before you get hurt.

"I am not a web monkey," you say. It is great that you have been writing software and database code for the last twenty years. Most of that will render little value to you in the web world. Sorry, but you are about to receive the "Web Whack Headache". Everything you currently know is all wrong for the web.

Finally, to my fellow Nerds who aim to impress others: Yes, you who add credential after credential to your e-mail signature and *résumé* until the acronyms and jargon rolls line after line across the entire page–this book was written for your kind.

TIP: Research title and acronym psychology to see if people *are* really impressed by the use of them before attaching them to your name.

To all others: If you think Visual C# is your goal, I believe you can do it. This is a reachable objective (claim is not guaranteed). Work hard, do not quit because quitters are losers who will *never* win. They can't because they never *finish* the race. Your first goal is to make it to the end of this read. It is not that long and if you cannot survive the pain of this book, then this Visual C# stuff is probably not your thing. It is okay if this flavor of coding turns out not to be your thing because is not for everyone, including those who already have achieved Nerd status. There is no shame in not attaining Visual C# credentials. Be who you are and do not try to imitate someone you are not.

A real winner runs the race and wins. To get there they practice endlessly before, during and after the race. They become good enough to outclass the competition. To win the C# race, you will need to become nerdier than you already are. Your benchmark is simply to have learned more today than you knew yesterday. Do not try to compare yourself to others. Focus on being your own benchmark.

Chapter 2

You are About to Write Code

Congratulations! You have made an important decision. You have decided to either keep reading this book, or to try Visual C#. Regardless of the direction you have chosen, you are about to embark on an important and slightly painful learning experience.

Now it is time to ask yourself: what type of code project will I pursue? You can go in a number of interesting directions with Visual C# because the Microsoft .NET Framework supports both web code and Windows application code (WinForms). Web code means you can build a website with Visual C#. Application code means you can build a software program that installs on a workstation. The advanced Nerd could also use application code to build a program that runs a mobile device, manipulates a data environment or integrates a system.

Web or application code? What is the better build direction to pursue? It all depends on the type of project you have in mind, and your requirements. There are pros and cons to both. See Figure 1 to check out out this list:

> **Help**
>
> Web = ASP.NET (supports C#)
>
> WinForm = Windows Forms, Win32, Win64, Windows API

Figure 1. *Acronyms Clarified.* Daniel Diaz, 2012.

Web - Pros

- A browser gets you to the cloud, which means low client maintenance.

- Code upgrades deploy on the server only.
- Security policies are flexible, with options in every environmental layer.
- Rapid scalability is an option; you can go from one to 10 MM users with ease.
- Web provides an efficient use of computing resources.
- If architected well, web code can be accessed by the average workstation consumer, as well as by mobile users, all working from one solution.

Web - Cons

- A network is required, such as an Intranet or the Internet.
- Web code is not good for heavy data processing.
- User interfaces are more challenging to build and less interactive than the alternative.
- A shared computing environment causes competition for server resources.
- The client / server environment is more complex to troubleshoot.
- There are limits to the computing tasks that can be performed via a web request.
- This approach requires knowledge of C#, HTML, CSS, browser behaviors, and client / server interaction.

App code - Pros

- You can personalize it for every need. You get the power of the clients computing environment dedicated to your data processing.
- You can harness the full power of personal computing. If you really need to kick off a thirty minute computing process, you can.

- There are more interface and usability options. This approach allows you to tap anything that happens in the computing environment.
- It is very flexible with many computing options for tasks. You can handle and run most tasks.
- Less knowledge is required to build the application–just Visual C#!
- The network detachment option is available. Take it on the road and sync later.
- Application code plays well with others. This approach is great for when you need solid, cross-platform options.

App code - Cons

- Deployment and upgrades are more challenging to manage. Every client needs your attention.
- Programmers don't always make it look good. Customers seem to think the web is better because it is prettier.
- Strong security can be challenging to achieve.
- You must consider multiple client OS environments.
- Application code will only run on the workstation. Alternative solutions need to be developed to run in other environments, such as a mobile device.

About GUI Tools

For you Nerds who have decided to disregard GUI coding tools: purchase Visual Studio in its many forms for your tool kit or simply use your favorite text editor and download the free Microsoft .NET compiler. GUI tools are not necessary, but they are strongly encouraged and seriously helpful. They offer a lot of benefits that you cannot get out of your freeware text editor that will increase your efficiency, and we all know that time is money. We can reserve the financial lesson for another occasion; in the mean time, check out some of the coding environment options we've listed below:

Tools for Web

- Visual Studio (paid)
- Visual Web Developer Express (free)
- Adobe Dreamweaver (paid)

Tools for WinForms

- Visual Studio (paid)
- Visual C# Express (free)
- SharpDevelop (free)

Improve your Experience

For the purposes of this book, we will focus in on what Visual Studio offers. This may offend some of you senior Nerds. Try to remember that we are keeping this book short and to the point. Examples and illustrations on every platform would eat up too much space. To help keep the content short and clear, illustrations are excluded except where complexity requires visualization. Remember that there will be minor differences between the Visual Studio version and Visual Express editions. If the mark is missed somewhere and you cannot find the option being described, then take a time-out and do a web search. The lingo remains much the same across these tools. It is simply a matter of finding the right option in your tool version and edition.

TIP: A real Nerd uses Internet searches for finding nearly everything Visual C# related.

TIP: As you code, you will be looking for the Namespace not the Library.

Keep in mind that Visual Studio has IntelliSense, which is a type of autocomplete tool. IntelliSense will help you form your code tags or display a list of object elements. This tool can be used in several ways. If you hover over code, you should see a Quick Info dialog appear with object information. Also, typing a tab, space, or period will display the Members List. Using Ctrl+Alt+Space will toggle the Completion Mode. Words can be completed using Ctrl+Space or Alt+Right Arrow.

TIP: There is something called the **Enterprise Library** provided on the MSDN website that is a collection of reusable software components. Check it out.

For you senior Nerds who feel ahead of the game and want more detailed information, keep in mind that this material is designed for the purpose of *learning*. It is not intended as a reference guide.

Chapter 3

Web Code Basics

Now it is time to learn about building a website or web based application. Here are some things you need to know. To start with the basics see Figure 2.

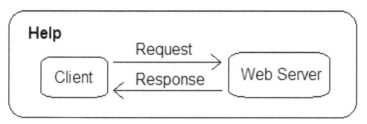

Figure 2. *Client/Server Interaction*. Daniel Diaz, 2012.

Web is client/server. The example (above) presents it in simple form. What you need to know about this interaction is that the exchanges are one-time events *initiated* by the *client*. For you software guys who have never worked with the web, this is the most important concept to absorb. It is present in *simple* form in the example provided. Seriously, these are *one-time* events. If you do not grasp this, then you are about to enter the headache zone. This zone is a place where additional learning may lead to physical injury. Take your time with this concept.

The fundamentals of web:

- The client is a web browser. JavaScript helps you work here.
- Your code is stored, processed, and served from a web server.
- The client needs a communication medium to talk with the web server. These can be through a LAN, WAN, or direct Internet connection.

13

- Client gets data via web protocols like HTTP, and HTTPS. Data is requested by client, processed and sent from web server, then passed to client browser.
  - HTTP is port 80
  - HTTPS is port 443
    - Web Server is the Cloud
    - Client gets to the Cloud via a URL

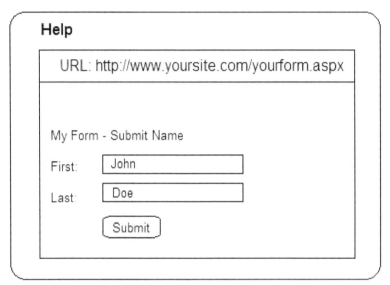

Figure 3. *Web Page.* Daniel Diaz, 2012.

You have used a web browser before, so let us look at a working example while considering Figure 3:

- Client initiates this process by visiting your web form (yourform.aspx) located on your web server (yoursite.com).
- When client first hits the page they should see empty form fields.

- Because client users are well trained web consuming monkeys they expect to enter their info (first and last name) then send it to the server via the [Submit] button.
- Once the server receives the client data via the Client Request, the server code will process the info, and then return a Server Response (i.e. some HTML output to provide interaction feedback like "yes it worked").

Take a look under the hood to see what is driving this HTML presentation. Look at this example:

```
...
<form method=post
action=process.aspx>
<label for=first>First:</label>
<input type=text id=first>

<label for=last>Last:</label>
<input type=text id=last>

<input type=submit value=Submit>
</form>
...
```

As the user types in "John" and "Doe" the browser is making a record of the information using the DOM. This info will remain with the browser until the client tells the browser to send it off to the web server via the [Submit] click or more technically the form POST.

When information moves from the web browser to the web server the data is being posted, at least in this example, to a script page called *process.aspx*. This receiving page can collect data via *GET* or *POST*. In this example *POST* was used because it is arguably the most common and simplest method to pass around web data.

TIP: You can search the Internet for HTML 101 to learn about some of basics of HTML and client/server interaction.

If a headache is starting to set in, then consider spending some time on *web fundamentals*. Start with very basic forms of HTML (i.e. .htm, .html). Once you have the HTML handled, consider understanding how scripting languages like ASP.NET, PHP, and Classic ASP use HTML. Learn how the power of scripting improves client/server interactions. Remember, the client (i.e. web browser) only understands HTML. Information that is returned from the web server needs to be in final form *HTML* otherwise the browser does not understand the message. A browser will do seemingly random and sometimes very funny things with non-compliance data. Now let us step this example up to ASP.NET:

The *example.aspx* code file (without a Master page)

```
...
<form runat=server>
  <asp:Label id=first Text=First
        runat=server/>
  <asp:TextBox ID=first
     runat=server></asp:TextBox>

  <asp:Label id=first Text=Last
        runat=server/>

  <asp:TextBox ID=last
     runat=server></asp:TextBox>

  <asp:Button ID=submit
     runat=server Text=Submit
     onclick=submit_Click />
</form>
...
```

TIP: A page with this form will post back to itself. Take note of the form tag <form runat=server> and how it has no specified post-to target. Also note how it implements the runat statement.

Explore the *example.aspx.cs* code file:

```
...
public partial class MyExample : System.Web.UI.Page
{
  protected void Page_Load(object sender, EventArgs e) { }
  protected void submit_Click(object sender, EventArgs e) { }
}
...
```

TIP: A code file is bound to the aspx page via the definer `Inherits`.

Here is a designer page example:

```
...
<%@ Page Language=C# AutoEventWireup=true  CodeFile=Default.aspx.cs
Inherits=YourClassName %>
<html>
<head runat=server>
   <title></title>
</head>
<body>
   <form id=form1 runat=server>
   </form>
</body>
</html>
...
```

TIP: You can switch to Single File Mode by using this page header format: <%@ Page Language=C# %><script runat=server>...</script>.

A code behind page example:

```
...
using System;
using System.Web;
using System.Web.UI;
using System.Web.UI.WebControls;

public partial class YourClassName : System.Web.UI.Page { }
...
```

Here is a Content page example using the Master Page model:

```
...
<%@ Page Title=MyTitle Language=C#
    MasterPageFile=~/MasterPage/
                MyPage.master
    AutoEventWireup=true
    CodeFile=MyCodeFile.aspx.cs
    Inherits=YourClassName %>

<asp:Content ID=Content1
    ContentPlaceHolderID=head
    Runat=Server>[Your head content]</asp:Content>

<asp:Content ID=Content2
    ContentPlaceHolderID=PlaceHolder
    Runat=Server>[Your page content]</asp:Content>
...
```

TIP: In Visual Studio, the Add New Item option will allow you to choose with or without Master Page.

In summary, your page will require two files such as the above example puts forth:

- *example.aspx*
- *example.aspx.cs*

Alternatively, you can follow another more complex coding pattern called Inline or Single File. This combines both concepts into one file. The distinction between these options will be the coder's preference. To keep our coding concepts clean, the two file model will be used throughout these examples.

Thanks for putting up with the web basics lesson for those who have this stuff down already. The web code wannabe's needed something to work from. Let's get into how C# and this .NET Framework stuff handles web. This will be a very high level perspective. If you are one of the physics textbook readers, then read away on the MSDN website because we will not be going into great detail. Senior Nerds consider searching for articles on ASP.NET Architecture. Look for the pictures with lots of arrows. ASP.NET is processed through the .NET Framework with its core intent to run on an IIS Web Server via a Microsoft Windows Server. See Figure 4 for the architecture model with a descriptive picture:

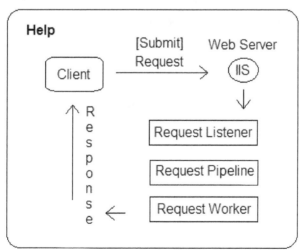

Figure 4. *Web Request Life Cycle.*
Daniel Diaz, 2012.

All you really need to know here is that the Microsoft .NET Framework runs inside of IIS (Web Server) to create a unique processing pathway for .NET related requests. The .NET Framework will in the simplest description look for .NET requests with the Request Listener. It will then move those web requests through the Pipeline. At which point a request receives a Worker who processes it then returns a Response. Remember the final Response will be in HTML for standard *aspx* style requests.

- Question: I am itching to code something, so how do I get a web server?

- Answer: Glad you asked because there are a couple different ways to achieve the server goal. The most expensive is buying a stand-alone server. A slightly cheaper option is to repurpose an existing one.

1) A cheap server path - To code, no server is required. Since this Visual C# stuff is ever evolving, inside Visual Studio is a shell IIS web server that can be deployed to test out and debug your code without the need for a real server. At the click of a button it will automatically launch your code inside a web browser that is pointed to this shell IIS instance. It is a very cool and much appreciated feature for this coding environment.

2) Getting public - You need a web server. I want to get past writing stuff on my workstation, so how do I make my stuff available to others? You can start out with a low cost option and there are many on the market.

TIP: This is very easily found by searching the web for .NET Windows Hosting.

Domain time - If you want to connect your website to a public domain name, most web hosts offer domain hosting options. If you do not find this option through your web host you can go to a domain host.

Though having a domain of your own can be fun to have, it is not needed if you plan to play in your own sandbox. A host provider will hook you up to a web server that has an IP address. That is all you really need to interact with a server. The domain is extra and can be added to make finding your server easy for the non-Nerds.

TIP: Learn more by searching the Internet for Domain Hosting.

Production time - You need FTP. I have my web server and I have my code so how do I get my code to the server? You will want to take advantage of a common methods of file transfer called FTP and SFTP. Your web server needs to have an FTP server to make either one of these options available to you.

- FTP is on port 21
- SFTP is on port 22

TIP: Search for Windows FTP Setup to learn about FTP options for your computing environment.

If you are on a LAN or WAN, then take advantage of native Windows file management methods. Look at sharing out a folder from your IIS Web Server. You will find the IIS installation defaults setup to a folder structure under file path C:\Inetpub\wwwroot.

TIP: Learn more about this topic by searching for Windows File Sharing.

TIP: Discover the real file path used by IIS by searching for Change IIS wwwroot.

Moving beyond the essentials, let's look more at what Microsoft .NET does so we understand what is happening as we write our code. The .NET Framework will take our scripted instructions, process them, and then render the final results into HTML form. This means all those fancy GUI objects like the ASP.NET Textbox on those `aspx` pages will go from object oriented style to HTML form as it goes through the framework translation process.

TIP: Before we get too technical, let's quickly talk about how to make a web project in Visual Studio. Open Visual Studio and click "New" and then "Web Project." The "New Project" window will open. Give it a name then get ready for the project to load.

TIP: To create a New Web Form go to the Visual Studio Solution Explorer and right-click the Website, and then select the Add New Item option. When the Add New Item dialog box is displayed chose the Web Form option.

TIP: What should you do if the Solution Explorer is not visible? This window usually disappears if you are one of those fast click don't know what just happen folks. Restore this window by View then Solution Explorer option or by key combo Ctrl+Alt+L.

If you have your coding environment open, go back to the HTML and look at the native form, then compare that to what ASP.NET will render in the browser. Do this by dragging a textfield to your form, compile, and then render it in a web browser.

Remember from an earlier example that the simple form for standard HTML textbox is done with the <input> tag. Check out what gets rendered through ASP.NET. It is the same thing but Microsoft .NET marks it up with all types of black box goodness to make the whole coding process easier (supposedly) and more manageable (possibly).

At the very least, .NET code is different. Whatever your take is, remember that it has become very popular because it gets most of the basic jobs done, and done well. If you are not sure how to see what HTML is rendered from your served web page then go to your web browser and find the View Source option. This will display the HTML your web browser received and processed.

TIP: In Mozilla view source by Web Developer > View Page Source option.

TIP: In Internet Explorer view source by Page > View Source option.

The ASP.NET web page has states and a life cycle. The provided chart, Table #1, is built based on the authors own assumptions. It is the page life cycle ordered and event driven events (this is real gold folks)

	Table 1 Web Page Life Cycle
1	System.Web.HttpApplication.BeginRequest
2	System.Web.HttpApplication.PreAuthenticateRequest
3	System.Web.HttpApplication.AuthenticateRequest
4	System.Web.HttpApplication.PostAuthenticateRequest
5	System.Web.HttpApplication.PreAuthorizeRequest
6	System.Web.HttpApplication.AuthorizeRequest
7	System.Web.HttpApplication.PostAuthorizeRequest
8	System.Web.HttpApplication.PreResolveRequestCache
9	System.Web.HttpApplication.ResolveRequestCache
10	System.Web.HttpApplication.PostResolveRequestCache
11	System.Web.HttpApplication.PreMapRequestHandler
12	System.Web.UI.Page.Construct

	Table 1 Continued **Web Page Life Cycle**
13	System.Web.HttpApplication.PostMapRequestHandler
14	System.Web.HttpApplication.PreAcquireRequestState
15	System.Web.HttpApplication.AcquireRequestState
16	System.Web.HttpApplication.PostAcquireRequestState
17	System.Web.HttpApplication.PreRequestHandlerExecute
18	System.Web.UI.Control.AddParsedSubObject
19	System.Web.UI.Control.CreateControlCollection
20	System.Web.UI.Control.AddedControl
21	System.Web.UI.Page.AddParsedSubObject
22	System.Web.UI.Page.AddedControl
23	System.Web.UI.Page.ResolveAdapter
24	System.Web.UI.Page.DeterminePostBackMode
25	System.Web.UI.Page.PreInit
26	System.Web.UI.Control.ResolveAdapter
27	System.Web.UI.Control.Init
28	System.Web.UI.Control.TrackViewState
29	System.Web.UI.Page.Init
30	System.Web.UI.Page.TrackViewState
31	System.Web.UI.Page.InitComplete
32	System.Web.UI.Page.LoadPageStateFromPersistenceMedium
33	Control.LoadViewState
34	System.Web.UI.Page.EnsureChildControls
35	System.Web.UI.Page.CreateChildControls
36	System.Web.UI.Page.PreLoad

	Table 1 Continued **Web Page Life Cycle**
37	System.Web.UI.Page.Load
38	System.Web.UI.Control.DataBind
39	System.Web.UI.Control.Load
40	System.Web.UI.Page.EnsureChildControls
41	System.Web.UI.Page.LoadComplete
42	System.Web.UI.Page.EnsureChildControls
43	System.Web.UI.Page.PreRender
44	System.Web.UI.Control.EnsureChildControls
45	System.Web.UI.Control.PreRender
46	System.Web.UI.Page.PreRenderComplete
47	System.Web.UI.Page.SaveViewState
48	System.Web.UI.Control.SaveViewState
49	System.Web.UI.Page.SaveViewState
50	System.Web.UI.Control.SaveViewState
51	System.Web.UI.Page.SavePageStateToPersistenceMedium
52	System.Web.UI.Page.SaveStateComplete
53	System.Web.UI.Page.CreateHtmlTextWriter
54	System.Web.UI.Page.RenderControl
55	System.Web.UI.Page.Render
56	System.Web.UI.Page.RenderChildren
57	System.Web.UI.Control.RenderControl
58	System.Web.UI.Page.VerifyRenderingInServerForm
59	System.Web.UI.Page.CreateHtmlTextWriter
60	System.Web.UI.Control.Unload

	Table 1 Continued Web Page Life Cycle
61	System.Web.UI.Control.Dispose
62	System.Web.UI.Page.Unload
63	System.Web.UI.Page.Dispose
64	System.Web.HttpApplication.PostRequestHandlerExecute
65	System.Web.HttpApplication.PreReleaseRequestState
66	System.Web.HttpApplication.ReleaseRequestState
67	System.Web.HttpApplication.PostReleaseRequestState
68	System.Web.HttpApplication.PreUpdateRequestCache
69	System.Web.HttpApplication.UpdateRequestCache
70	System.Web.HttpApplication.PostUpdateRequestCache
71	System.Web.HttpApplication.EndRequest
72	System.Web.HttpApplication.PreSendRequestHeaders
73	System.Web.HttpApplication.PreSendRequestContent

Source: GeekInterview:A Venture from Exosys.com
http://www.geekinterview.com/question_details/22673. 05/24/2012.
C# 411. http://msdn.microsoft.com/en-us/library/ms178472.aspx. 05/24/2012.

Stuff to Remember about the Web

- The coder is building served content.
- The clients are consuming HTML.
- Clients want interaction.

See Table #2 below for a list of frequently used events with actual class name info:

Table 2 Frequently Used Page Events	
LoadViewState	ViewState load event handler
LoadPostData	Postback load event handler
Page_Init	Page Initialization event handler
Page_Load	Page Load event handler
Page_PreRender	Page Pre Render event handler

Source: GeekInterview.com http://www.geekinterview.com/question_details/22673
5/14/2012.

What are these events? Basically, these are the page events you will need to hook into to do stuff with your ASP.NET application. Using the previous John Doe form example, when the client clicks that [Submit] button, it is these page events that drive what happens next. There will be a *Submit_click* event that will drive what happens once you have your client information.

TIP: Controls like the ASP Textbox and the ASP Button are derived from *System.Web.UI.Control.*

Wondering how to work with these events using Visual C# code? Check out these examples (more gold):

```
protected override void
LoadViewState(object savedState) { }

public virtual LoadPostData(string postDataKey, NameValueCollection
postCollection) { }
```

TIP: Implement the *IPostBackDataHandler* interface to use *LoadPostData*.

```
protected void Page_Init (Object sender, EventArgs e) { }

protected void Page_Load (Object sender, EventArgs e) { }

private void Page_PreRender(object sender, System.EventArgs e) { }
```

TIP: Don't forget to manage your event properly with *System.EventHandler(this. Page_PreRender)* when using *Page_PreRender*.

```
protected internal override void
Render(HtmlTextWriter writer) { }

protected void Page_Unload (Object sender, EventArgs e) { }

public virtual void RaisePostDataChangedEvent(){ }
```

TIP: Implement the *IPostBackDataHandler* interface to use *RaisePostDataChangedEvent*.

```
public void RaisePostBackEvent(string eventArgument){ }
```

TIP: Implement the *IPostBackDataHandler* interface to use *RaisePostBackEvent*.

TIP: Research *AutoEventWireup*.

Master Pages

Think of the Master page as a template tool. Its intent was to help you create a standard look and feel for all of your application pages. A Master page provides one place to manage your standard content such as the Application Header, Application Footer, Application Navigation, and interface structure. In .NET language, your "other" pages are considered Content pages when you use the Master page model.

TIP: When creating a new Content page, select the "use a Master page" option at creation time. The wizard will help you find the pre-created Master page to link to your new Content page. If you missed this step, you can always code the link yourself in the Content page header. Just follow the expected Master/Content code structure.

[1][2]Here is the Master page life cycle (more gold):

1. Content page PreInit event.
2. Master page controls Init event.
3. Content controls Init event.
4. Master page Init event.
5. Content page Init event.
6. Content page Load event.
7. Master page Load event.
8. Master page controls Load event.
9. Content page controls Load event.
10. Content page PreRender event.
11. Master page PreRender event.
12. Master page controls PreRender event.
13. Content page controls PreRender event.
14. Master page controls Unload event.
15. Content page controls Unload event.
16. Master page Unload event.
17. Content page Unload event.

Meeting the PostBack Event

We will spend a little time on the *PostBack* because it is a frequently used event. If you are about to build anything substantial, you will probably use it.

Let's understand when it happens. The first time a page is loaded you can check this value *Page.IsPostBack*. It will read *false*. However, if we fill out a form and click a [Submit] button, the page should post back to itself. Check the value again to verify that it is now changed to *true*. This is useful if you want to manage the state of your page. It is also useful if you need a hook to evaluate if this is the first time your page has been loaded by the user. A common use of this *Page.IsPostBack* is in the *Page_Load* event handler. Here is an example:

```
using System;
using System.Web;
using System.Web.UI;
using System.Web.UI.WebControls;

public partial class _Default : System.Web.UI.Page {

protected void Page_Load(object sender, EventArgs e) {

bool MyValCheck = Page.IsPostBack; } }
```

FAQ: There is a sequence of events raised for Pages, UserControls, MasterPages, and HttpModules. Understanding the page life cycle can be very important as you begin to build pages using a mix of these elements. Do not take this as gospel for what the order of these events is or should be. You can test them out for yourself so that you *know* when these events fire and in what order. Do this so that your application runs as you expect. Test them when the value is this *postback=false*, and when the value is that *postback=true*. Check through any other page states you may have fabricated for your application.

Web.config

The *web.config* file is an XML document where configuration information is stored including a place for your own custom settings.

Working file example:

```
<?xml version="1.0" encoding="utf-8" ?>
<configuration>
<system.web></system.web>
<appSettings>
<add key="example" value="some_data" />
</appSettings>
</configuration>
```

Reading from the file example:

*MyValue=System.Configuration.ConfigurationManager.AppSettings["**example**"]*

ViewState

The purpose of *ViewState* is as a state manager for postbacks. ViewState data is sent to the client. Then the stored data is serialized and sent back to the server via a hidden input field. You need to be aware of this extra data being stored on your pages because the more it is used the more your page size can increase. The resulting change will impact page performance. You can find the ViewState field by searching in your *aspx* file for *id="__VIEWSTATE*.

TIP: ViewState usage is a good area to work on to improve page performance. Page performance and page size (amount of data transferred from server to client) is another dimension to consider in your development process. Here size matters!

You can control whether ViewState is used in your application by toggling its use via the *EnableViewState* option.

ViewState example:

> *<CONTRL EnableViewState="value"*
> (accepts *true* or *false*)...

Global.aspx

Keeping this simple, the *Global.aspx* is the place to manage application level configurations. Here is an example file outlining some of the features available at this global access level:

```
<%@ Application Language="C#" %>
<script runat="server">
void Application_Start(object sender, EventArgs e) { }
void Application_End(object sender, EventArgs e) { }
void Application_Error(object sender, EventArgs e) { }
void Session_Start(object sender, EventArgs e) { }
void Session_End(object sender, EventArgs e) { }
void Application_EndRequest(object sender, System.EventArgs e) { }
</script>
```

All this is ground work and we have not even mentioned good stuff like CSS, Ajax, Business Logic, and Databases.

Chapter 4

Application Code Basics

It's time to build some type of Windows application that can be installed on a workstation. Where in, no servers are required, but can be used if needed. The door is open to many things that can be done with this code base. But keep in mind application code is different from web stuff. If you try building a web form using the WinForm as the GUI it will not work so well once it is on the web server. Browsers will have an allergic reaction to seeing an .exe file. The browser assumes it to be bad because this is one of the ways to transmit a computer virus.

TIP: Before we get too technical, let's quickly talk about how to make a Windows Application project in Visual Studio. Open Visual Studio and click "New" and then "Project." Your looking for a Visual C# Windows project. Once you find it and the wizard loads give your project a name. Finally, wait for the Visual C# Windows project to load in the IDE.

TIP: To create a Form, go to the Visual Studio Solution Explorer, and right-click the Solution header and then select the Add New Item option. The Add New Item dialog box is displayed, choose the New Form option. Follow these instructions if the Solution Explorer is not visible. This would also be helpful if you happen to click too fast making the windows disappear. This window can be restored from the File Menu's View option and then select the Solution Explorer choice or by keying a combination of Ctrl+Alt+L.

TIP: Learn to toggle between Code view and Designer view. If you are in Designer view, you can switch to Code view by right-clicking in the Designer window and then clicking View Designer and vise-versa.

TIP: Inside Visual Studio environment you should be able to see a Toolbox menu that lists out most of the form controls available for Windows application development. Try the shortcut Ctrl+Alt+X to get this Toolbox to be displayed.

TIP: The final window you want to be familiar with is the Properties Window (F4). Usually you can get this to appear by right-clicking on any visual object in your design screen then choosing the Properties option.

Through the use of API functions and .NET Framework elements you can access operating system features like:

- Directory Services
- Group Policy
- Computer Health
- Computer Manager
- Task Scheduler
- Windows Deployment Services
- Windows Remote Management
- Windows Security

TIP: Looking for more on API? Check out the Windows Template Library and the Microsoft Foundation Classes.

Fundamentals of Windows Applications

The client can run on a workstation. Code can be compiled, processed and then served from the client. Unlike web code, no external communication mediums are required to provide access to the application. We can decide how the client gets data when it is needed. Data can be delivered via web protocols such as HTTP and FTP. It can also be through Windows File protocols using direct file locations such as (C:\) or UNC (\\Server\SharedFolder\) paths. It can also be through Web Services or direct database system communication.

Similar to what ASP.NET delivers in terms of event driven programming we have similar access to events inside a Microsoft Windows application. Here are a couple of event examples:

```
private void MyForm_Load(object sender, EventArgs e) { }
private void MyForm_Closing(object
sender,System.ComponentModel.CancelEventArgs e) { }
private void MyForm_FormClosed(object sender, FormClosedEventArgs e) { }
```

The Form Show/Load/Close events with indication of visibility is through *System.Windows.Forms*:

Table 3 Show and Close Events	Visibility
MyForm.**Show**	No (Starts Load)
Control.HandleCreated	No
Control.BindingContextChanged	Yes
Form.**Load**	Yes
Control.VisibleChanged	Yes
Control.GotFocus	Yes
Form.Activated	Yes
Form.Shown	Yes
MyForm.**Close**	Yes
Form.Closing	Yes
Form.Closed	Yes
Form.FormClosed	Yes

Table 3 Continued Show and Close Events	Visibility
Form.Deactivate	Yes
Control.HandleDestroyed	Yes
Component.Disposed	No

Source: *C# 41.* http://www.csharp411.com/c-winforms-form-eventorder/1. 5/14/2012.

Form Show/Hide events with indication of Visibility through *System.Windows.Forms:*

Table 4 Visible Events	Visibility
MyForm.Visible = false	Yes (Hide Proc)
Form.Deactivate	Yes
Control.LostFocus	Yes
Control.VisibleChanged	No
MyForm.Visible = true	No (Show Proc)
Control.VisibleChanged	Yes
Control.GotFocus	Yes
Form.Activated	Yes

Source: *Microsoft Developer Network.* http://msdn.microsoft.com/en-us/library/86faxx0d.aspx 5/14/2012.
C# 411. http://www.csharp411.com/c-winforms-form-event-order/ 5/14/2012.

Form events traced (more gold):

	Table 5 Form Event Order
1	System.Windows.Forms.ClientSizeChange
2	System.Windows.Forms.ControlAdded
3	System.Windows.Forms.Constructor
4	System.Windows.Forms.HandleCreated
5	System.Windows.Forms.Invalidated
6	System.Windows.Forms.Load
7	System.Windows.Forms.Loaded
8	System.Windows.Forms.CreateControl
9	System.Windows.Forms.OnActivated
10	System.Windows.Forms.Shown
11	System.Windows.Forms.OnPaint
12	System.Windows.Forms.Invalidated
13	System.Windows.Forms.OnPaint

Source: *stackoverflow.* http://stackoverflow.com/questions/397121/winforms-load-vs-shown-events/397810#397810
5/14/2012.

TIP: Performing your own application testing is important. Disclaimer: This data set should be taken with a grain of salt. As suggested in a previous section, test your code to make sure things operate as expected. This info is provided as a reference of what to look for.

Now let's start to put these pieces of information together to work towards building something. Windows Forms can be designed around events like *Form_Load*. Events provide a hook to the action and reaction of user and system input and output. Through these hooks you can, in a very detailed way, architect your application interactions and processing.

To get your project code on your computer to another workstation look into building a Windows Installer project. This type of project will hook into your application code and build outputs. It will build them into a simple installation process that can be deployed to a shared storage device, network location, or burned to physical media. If the options of a Windows Installer do not do the job, then check into some popular third party installer options.

Application Planning Considerations

- Application security (user access and authentication vehicles).
- Operating system security (who can execute the app, what privileges does the app require).
- Screen display (compatibility for users who don't use the same screen resolution nor have the same resolution ratios – wide vs. standard etc.).
- Application performance (hard drive, memory, cpu, processes priority, graphic usage etc.).
- Multiprocessing will use threads in Windows.
- Operating system and platform versions (clients might be running different clients or the same clients but with different patches installed).
- Conflicts with other software and standard business systems (ex. Firewall, Antivirus, Device Driver etc.).
- Globalization (language support).
- Application architecture.
- System architecture.

TIP: Do not try reinventing the wheel. There is a good chance if you are stuck on something; someone has already built some type of solution to deal with a problem. Check into third party options along with online help guides.

Finally, another word of advice, if you are going to build something for real, make it look decent. Put some care and thought into your GUI design and interactive experience. Make it look and feel cool and your client may fall in love with it. You are offering something visual. Granted this is virtual visual, never the less, it is the look that offers the vibe of usability and quality.

Build Details

There is a class file that can be added to your project called *AssemblyInfo*. It provides build details such as; build version information that gets compiled into your application output. This file will hold other informative details as well. Browse through this file to get an idea of what you can do with it.

TIP: If you want to build for other platforms search for Compile C# [platform_name]. Check out what is good for Linux, Mac etc.

TIP: Look for the InternalsToVisibleAttribute that is address in the later *Testing Time* chapter.

Chapter 5

Think Usability

Computer human interactions are all about Psychology. Humans behave in certain ways and they have certain expectations. Usability results from understanding human psychology and working within those expectations. Unfortunately, technology is not always able to keep pace with our human expectations and desires. Sometimes we have to push technology as far as it will reach and then adjust our human expectations about how much technology will realistically be able to help.

This is a good opportunity to create some awareness about the tools provided in your software development toolkit. The goal is to give you something to work with as you build your human-computer interactions with both expectations and limitations in mind.

As you design and build applications and computer systems, take time to evaluate whether you have achieved usability. This usability will be defined as accepted human computer interactions. This means us humans tolerate using the computer interface. Use of the computer interface and toleration of such can be tested. The basic test process looks like:

1. Design test
2. Run test on real humans, (typical individuals - not Nerds)
3. Analyze test results
4. Make changes to identified areas
5. Repeat process

At the end of the refinement process, you should have an interactive tool that humans can tolerate. If you do a really good job, humans may even accept and embrace use of the end product, achieving usability.

Testing usability is a simple concept and can be evaluated by doing the following:

- Place the new computer tool before a human
- Inform tester of the tool purpose and function
- Ask individual to perform a specific task
- Observe their activities
- Get their feedback

When you observe your testing subjects, see if:

- The task is performed efficiently
- The subjects can find their way to specific functions and tasks
- The subjects can and will provide constructive feedback

Tolerated usability may have been achieved when a task is done without too much effort exhausted in figuring out how to do something without undue stress. You want your software to avoid obvious problems that cause the tester to undergo stress resulting in unnecessary hair loss or the development of the need to throw things. We are looking for specific functions and tasks to be completed while maintaining a calm demeanor.

At the end of the task do not forget to collect feedback. The feedback you receive while watching them perform the task and the feedback you collect from their direct communication is what you can work with to fix and enhance the interactive experience. At this point it is okay to collect more feedback from other Nerds. Just remember who the end customer really is and what input they have provided.

TIP: When you need more help consider seeking feedback and consultations from technological professionals. They can help you improve your system. Bear in mind that you are the one likely responsible for making sure their improvements are properly aligned with the end user.

Design with Purpose

Some people feel compelled to build things for no reason. Others build because they are trying to learn something new. However, in a business context, no one wants an application that looks cool, and does cool stuff if it does not serve some real function and purpose. The function or purpose of your application should be obvious. This is what we refer to as Purpose Driven Programming. Build with some intent or you are just wasting yours and others time, which violates the first nerd law on user interface design. We cannot have that. Make your application build purpose clear. People should know what it does without having to ask you or another Nerd.

TIP: No kidding on the Nerd law chatter. We really have them so take the time to find them, and learn them. Nerd laws are like "man laws" except they are expressed because they were developed by and for Nerds.

Basics of Visual Design

Put some care and time into your interface designs. Your goals are to have your designs:

- Be visually appealing
- Be easy to use
- Have a logical flow
- Be intuitive for use and locating items

You want your final product to be visually appealing, easy to use, and consist of a logical flow. The goal is to make it intuitive for the user to find things and use them.

Consider design and layout in terms of number of flicks or steps to execute the various operations available. Ask yourself:

- "How many mouse clicks will it take for the user to go from this screen to the next one?"
- "How many clicks will it take the user to work through this section?"

Make your operations concise. Summarize your operation options into clean, easy to follow patterns. Pare your descriptions down to seven words or less. Finally, get design feedback before your end customer sees it. Be honest with yourself about what you have. Always work within your means and your budget.

Make your GUI design simple. We Nerds make things more complex than they really need to be. It requires real know-how and creativity to take something complex like this computer-human interaction stuff and turn it into something simple. We are talking simplicity that humans not just tolerate but something they may actually like. Here is a secret- companies like Apple, Microsoft, IBM all found a way to build software people *like* to use. Figure out how to do this and you might actually be successful. Do a good enough job and you might even impress other Nerds.

TIP: There is an old Nerd saying: Keep it simple stupid. Official Nerds beg you to follow it.

Basics of Performance

People (aka non-Nerds) use software all the time. They have certain performance expectations. Try hard to meet their expectations. Bend your application around their views of how software should work. Start with keeping under 2-3 seconds for a screen load. Consider providing some type of feedback to the user if something is going to take longer than they think it should.

TIP: Search for User Expectations

Basics of Reusability

Always consider how your code can be reused and repurposed. Build for reuse. This concept is why senior Nerds build elegant frameworks around code. They build using DLL's, classes, interfaces, and use this stuff called enterprise architecture. It is done for *reusability*. If you are going to build it, try to build it just <u>once</u>.

TIP: The smart Nerds follow the "be lazy" coding principal - code once for reuse. It is not a Nerd law yet.

C# Style Guides

There are many style guides that have been written, over the years specific to different code platforms. The author is not going to recommend any one over another but will suggest that you want to use a guideline that will lend to coding consistency. Work from rules that are easy to follow and implement. Use a process that makes written code understandable to other Nerds.

TIP: Search for C# Style Guides

Web Controls

You can drag and drop web controls from the Visual Studio toolbar onto your web form or you can write them into your *.aspx* code file. The drag and drop option is nice and quick. However, if you really get into a loaded interface, you may find it necessary to code manually.

Standard Controls:

- AdRotator
- BulletedList
- Button
- Calendar
- CheckBox
- CheckBoxList
- ContentPlaceholder
- DropDownList
- FileUpload
- HiddenField
- Hyperlink
- Image
- ImageButton
- ImageMap
- Label
- LinkButton
- ListBox
- Literal
- Localize
- MultiView
- Panel
- PlaceHolder
- RadioButton
- RadioButtonList

- Substitution
- Table
- TextBox
- View
- Wizard
- XML
- Data Controls:
- AccessDataSource
- DataList
- DataPager
- DetailsView
- FormView
- GridView
- ListView
- Repeater
- SiteMapDataSource
- SqlDataSource
- XmlDataSource

Validation Controls:

- CompareValidator
- CustomValidator
- RangeValidator
- RegularExpressionValidator
- RequiredFieldValidator
- ValidationSummary

Navigation Controls:

- TreeView
- SiteMapPath
- Menu

Login Controls:

- ChangePassword
- CreateUserWizard
- Login
- LoginName
- LoginStatus
- LoginView
- PasswordRecovery

Webpart Controls:

- AppearanceEditorPart
- BehaviorEditorPart
- CatalogZone
- ConnectionsZone
- DeclarativeCatalogPart
- EditorZone
- ImportCatalogPart
- LayoutEditorPart
- PageCatalogPart
- PropertyGridEditorPart
- ProxyWebPartManager
- WebPartManager
- WebPartZone

Ajax and Available Options:

- ScriptManager
- Timer
- UpdateProgress
- ScriptManagerProxy

- UpdatePanel

Custom Control:

If an existing control does not meet all your basic requirements, you can extend it, or tailor it to meet your needs. There are several different ways to build and utilize custom controls.

Windows Application Controls

These controls can be pulled on to your form via drag and drop just like you can with web controls from the Visual Studio toolbar. IntelliSense helps keep your code well formed, and highlights the many options you have available.

Standard controls:

- BackgroundWorker Component
- BindingNavigator
- BindingSource Component
- Button
- CheckBox
- CheckedListBox
- ColorDialog Component
- ComboBox
- FileDialog Class
- ContextMenu Component
- ContextMenuStrip
- DataGrid
- DataGridView
- DateTimePicker
- Dialog-Box Controls and Components
- DomainUpDown

- ErrorProvider Component
- FlowLayoutPanel
- FolderBrowserDialog Component
- FontDialog Component
- GroupBox
- HelpProvider Component
- HScrollBar's
- VScrollBar's
- ImageList Component
- Label
- LinkLabel
- ListBox
- ListView
- MainMenu Component
- MaskedTextBox
- MenuStrip
- MonthCalendar
- NotifyIcon Component
- NumericUpDown
- OpenFileDialog Component
- PageSetupDialog Component
- Panel
- PictureBox
- PrintDialog Component
- PrintDocument Component
- PrintPreviewControl
- PrintPreviewDialog
- ProgressBar
- RadioButton
- RichTextBox
- SaveFileDialog Component
- SoundPlayer Class
- SplitContainer

- Splitter
- StatusBar
- StatusStrip
- TabControl
- TableLayoutPanel
- TextBox
- Timer Component
- ToolBar
- ToolStrip
- ToolStripContainer
- ToolStripPanel
- ToolStripProgressBar
- ToolStripStatusLabel
- ToolTip Component
- TrackBar
- TreeView
- WebBrowser

Custom control:

If an existing control does not meet all your needs, you can extend it, or just build your own control. There are several viable options to build and employ custom controls. If you need to customize, take the time to research out your options.

TIP: Search the web for control name to see use examples. Do a little reading on other Nerd experiences and you might pick up on each controls features and failure points.

Other .NET Projects:

- Business Intelligence
- Analysis Services Project
- Integration Services Connection Project
- Integration Services Project
- Report Server Project
- Report Model Project
- Microsoft Office Add-in
- Office Document Project
- Office Template Project
- WPF Browser Application
- WPF Custom Control Library
- WPF User Control Library
- Windows Forms Control Library
- Class Library
- WPF Application
- WCF Service Application
- Console Application
- Smart Device Project
- SSIS and Sql Server Projects
- Reports Application
- Crystal Reports Application
- Test Project
- Sequential Workflow Service Library
- Syndication Service Library
- State Machine Workflow Service Library
- WCF Service Library
- Sequential Workflow Console Application
- Sequential Workflow Library
- SharePoint Sequential Workflow
- SharePoint State Machine Workflow
- State Machine Workflow Library
- State Machine Workflow Console Application

- Workflow Activity Library
- Application Design
- System Design
- Logical Datacenter Design
- Setup Project
- Merge Module Project
- CAB Project
- Setup Wizard
- Smart Device CAB Project
- Visual Studio Add-in
- Shared Add-in
- Web Service

Chapter 6

Know How and Where to Put the Rules

Now you are going to start to build the meat of your project. If you need rules, keep reading so you know where to put them, and how to form them. These business logic rules are good to keep packaged together, but some software architectural design patterns call for an application layer that handles business logic. These layers are good to build on frequently used coding elements.

There are several components that are both frequently used and critical for awareness:

- Classes
- Interfaces
- Structs

In order to build code using these tools, it's important to understand:

- Methods
- Properties
- Variables and Operators
- Data types
- Security controls
- Generics
- LINQ

TIP: You will not receive many spelled out examples for this information. Now that you know what to search for, use the Internet as a search tool.

Classes

You can build class files that are separate from your interface functionality. They can run and hold your business logic. Under object oriented coding patterns, this logic would find its way inside a class.

Possible members of classes:

- Fields
- Constants
- Properties
- Methods
- Constructors
- Destructors
- Events
- Indexers
- Operators
- Nested Types

Classes in Visual C# can be declared as static or left without declaration. A non-static class will result in a separate instance being created when used. A static class will represent one shared copy in system memory.

A class in Visual C# can only inherit from one base class. Also, when a class uses a base class, it inherits all the elements of the base class except the constructor section. In our example below Child is known as a derived class.

```
public class Parent
{ public Parent() { ... }
public void Example() {...}  }

public class Child:Parent
{ public Child() {...}
```

```
public static void Main()
{ Child child = new Child();
child.Example(); } }
```

Interfaces

Interfaces describe a group of related functionalities that can belong to any class or struct. An interface can consist of methods, properties, events, and indexers. You can have multiple interface inheritance. Here is an example using the interface:

```
interface IParent { void Example(); }

class Parent:IParent {
public void Example() {...} }
```

Structs

A struct type is a value type that can contain constructors, constants, fields, methods, properties, indexers, operators, events, and nested types. They are custom values types. They also do not require heap allocation which makes them a good choice for improving performance.

Possible members of structs:

- Fields
- Constants
- Properties
- Methods
- Constructors
- Destructors
- Events
- Indexers

- Operators
- Nested Types

Here's a working example of a Struct:

```
public struct Name {
public string First { get; set; }
public string Last { get; set; } }

class Customer {

void ProcessPostback()
{ Name name = new Name();
name.First = "John";
name.Last = "Doe";
Save(name); }

void Save(Name name) {...} } }
```

The Tools

What is a method? A method is a block of code that is capable of both receiving and returning values.

What are properties? Properties are a way to read, write, and/or do computations. Properties are also known as accessors. They are usually distinguished by the *get* or *set* key words.

What are variables? Okay, let's be real for a moment. If you are reading this book and you do not know what a variable is, you should be spend the time to educate yourself on that concept with other resources. For everyone else, C# is type safe, therefore we have several variable forms to choose from: static, instance, array, value, reference, output, and local.

TIP: Need operator options? Search for C# Operators.

Frequently used data types:

- bool
- byte
- char
- decimal
- double
- float
- int
- long
- object
- string

Security Controls: Also known as access modifiers, security controls have four main choices to work with: **public** (open door), **private** (closed door), **protected** (class only), and **internal** (assembly only).

Generics: Think of this topic as extended data types. This model allows you to softly change the standard behavior of data types without abandoning them.

LINQ: Also known as Language-Integrated Query, allows you to write queries detached from a specific data source. Note that this is not a replacement to understanding how database systems work. Take the time to understand when to use and when not to use LINQ. This is not a replacement for SQL. If you are going to pull lots of records from a database and then transmit them to your application just to filter the records down further, consider this. It is still prudent to use SQL on your database to filter to the records you really intended to use before pulling this information into your application. Your users want interaction and, not a spinning hour glass. With that said, LINQ is a good option to express filters in the form of a query. They are great for readability.

TIP: If you are concerned about system performance, employ a divide and conquer strategy by moving the heavy processing to other less used servers. Also, think Threads.

Chapter 7

Put Data in the Driver's Seat

Welcome to the world of ADO.NET. ADO.NET is a technology platform designed for interacting with data sources. ADO.NET has specific vehicles for direct interaction with specific database systems. However, there are other options available besides ADO.NET. If you get bored with ADO.NET, or simply need to use a different communication vehicle, you can always choose from a different .NET Data Provider option such as ODBC and OLE DB.

What you Need to Know

- ADO.NET has purpose;
- MS SQL is native;
- MySQL is available but not native;
- Oracle is available but not native; and
- XML is native, but be warned that it is not always standards compliant.

If you want to work with Microsoft SQL database systems, check out the SQL Data Provider. In ADO.NET you need to know that these things exist:

- SqlConnection (your connection definition)
- SqlCommand (db communication gateway)
- SqlDataReader (efficient for reading data)
- DataSet (holds data in memory)
- SqlDataAdapter (data management tool)

If you want to work with Oracle database systems, check out the Oracle Data Provider. There are similar objects to Microsoft SQL for

Oracle. Users can expect a similar object structure and interaction mechanisms to that they would find in Microsoft SQL.

If you want to work with MySQL database systems, check ADO.NET's Driver for MySQL. Like the other objects discussed above, MySQL has a similar arrangement and interface to the objects you're accustomed to using.

TIP: Microsoft Office tools, like Access and Excel, can be queried via the Microsoft Jet ODBC Driver and the Microsoft ODBC Desktop Database Driver.

TIP: ADO.NET is all about data. If you do not have a level of familiarity with databases, it's advisable to get used to using them before writing code around them. Check out some SQL 101 to help get you up to speed.

SqlClient Sample (*System.Data.SqlClient*)

```
conn = "your string"
sql_qry = "SELECT stuff"
using (SqlConnection sqlconn = new SqlConnection(conn)) {
SqlCommand cmd = new SqlCommand(sql_qry, sqlconn);
// use something to pull the data out like a SqlDataReader }
```

TIP: For these examples, it is assumed you know what a connection string looks like. If you don't know what one looks like, search for "connection strings". There is some real magic in the formation of the string. Ensure you have the right format and version for the type of database you are connecting too.

OleDb Sample (*System.Data.OleDb*)

```
conn = "your string"
sql_qry = "SELECT stuff"
using (OleDbConnection oleconn = new OleDbConnection (conn)) {
OleDbCommand cmd = new OleDbCommand (sql_qry, oleconn);
// use something to pull the data out like a OleDbDataReader }
```

Odbc Sample (*System.Data.Odbc*)

```
conn = "your string"
sql_qry = "SELECT stuff"
using (OdbcConnection odbcconn = new OdbcCommand (conn)) {
OdbcCommand cmd = new OdbcCommand (sql_qry, odbcconn);
// use something to pull the data out like a OdbcDataReader }
```

TIP: Hopefully, the code to this point has been simplified enough to make you think this book is da-bomb. If not, then search from more info on the specific objects addressed.

ADO.NET Web Service (*System.Web.Service*)

With Visual C# you are able to build and consume web services. Web Services are a vehicle for independent platforms to exchange and share data.

Web Service example (*System.Web.Services*):

```
<%@ WebService Language=C# Class=Exp %>
using System.Web.Services;
[WebService(...)]
public class Exp : WebService
{ [WebMethod(...)]
public string Example() { return } }
```

Consume Service example:

```
...
Exp MyExample = new Exp();
string data = MyExample.Example();
...
```

Through ADO.NET, you can return objects such as a DataSet through your web service. However, if you use this feature, it somewhat defeats the point of the Web Service. Remember the primary use of the Web Service is a data exchange for platform independence. Therefore it is worth considering keeping your Web Service exchanging primative typed XML unless you are certain .NET is the only way you will be consuming.

Chapter 8

Learn HTML First

This section applies to ASP.NET coding. It is for those who want to whole heartedly build web forms. Things web Nerds really need to know are:

- HTML
- JavaScript
- CSS

Why do I need to know these things you might ask? The answer is simple. If you have no idea what is happening on the backend, you will have an impossible time trying to troubleshoot a problem.

Once the app is broken, people tend to lean to pretending it is not. Eventually, one must reason with reality, which usually results in calling in the big guns. That call is usually to a senior Nerd, or to the Nerd familiar with physics books. To be frank, the reason for this book is from repeatedly fixing ignorant problems that could have been avoided by understanding the basics. If reason can be spread before an ignorant problem develops, there is a chance at working in some rational thought before the problem gets irrationally out of control. This dissatisfaction comes from having to tell people it will cost 2MM to fix it but the whole thing can rebuild it from scratch for 1M. Never mind that someone has burned 3MM on it so far. This usually results in the realization that the big guns should have been hired from the start. Do not become the cause for burning other people's money. They usually don't like that and it results in very upset customers. This action is not good for job security of amateurs, but it is usually good for those like me. Maybe the big guns should offer thanks in advance for any forthcoming destruction that may come from the novice, and scrap the book so that no one learns to get it right before the project starts. Enough ranting, let's look at some of these basics.

Start with HTML. This is a markup language, it is frequently used as the basic building blocks for web. It is similar to XML but with its own tags, elements, and rules. Try to make your final code fall into compliance with one of these standards:

- W3C HTML 4.01
- W3C HTML 5

TIP: Search for HTML 101 to learn more.

TIP: Search for HTML Validator to write standard compliant code.

JavaScript (JS) is a client side scripting language that allows you to extend outside the parameters of HTML and enhance the interactive experience. It is useful for making interactions more dynamic and is the basis from which Ajax frameworks are built upon. Most importantly JS provides access to the DOM allowing you to manipulate the *client* environment.

TIP: For advanced learning, check into the Document Object Model (DOM). JavaScript, Ajax, and JQuery all give access to the DOM.

TIP: Write standard compliant JavaScript by searching for JavaScript Validator.

TIP: Consider using a third party JavaScript debugging tool. Get one by searching for JavaScript debugger.

Cascading Style Sheets (CSS) can be considered a style framework for web presentation. Usually applied to HTML and XHTML, it does have other less common uses, but we will not go into those here. In this case you can integrate it with your ASP.NET application. CSS styling will cover things like: font color, font size, font type, background color, border style, position of element, object padding etc. The Visual Studio environment provides access to easy CSS application with highlighted value place holders. Consider having your final CSS code fall into compliance with one of these standards:

- W3C CSS 1.0
- W3C CSS 2.1

TIP: Search for CSS 101 to learn more about CSS.

TIP: Search for CSS Validator to write standard compliant code.

TIP: Add a CSS file to your project by Solution Explorer > Right Click > Add New Item > Style Sheet.

There are now many plugins available for web development that aids the coding and troubleshooting process. Tools like Firebug and Web Developer will help take your game to the next level. The more information you can get from your web browser the less pain you will feel when you hit a coding snag.

Before you start, know your audience. Make sure you are coding to the browser type and version your clients will be using. Some environments can be controlled so it is acceptable to write to the controlled environment. Some environments like public Internet sites should be written to whatever browsers are in popular use. It is not fun to write code for several browsers but the more you handle the better the experience for others, and the more your work will be appreciated, increasing your business and career options.

TIP: Search for Web Browser Statistics to get the most up-to-date view of general browser utilization.

In addition to smaller coding tools and add-ins, as mentioned already, there are things built into the Visual Studio environment to help you with your code. These aids will help you identify compliance issues. Additionally, you can get add-ins for your Visual Studio environment that will enhance coding even more. Also, you can find many other web developer toolkits and software aids available. Do not be afraid to take the time to know what other tools are at your disposal. Be open to all the resources available to you.

TIP: Search for Visual Studio Add-in to get an idea of other features you can add to your coding environment.

Chapter 9

The Framework Has It

There are many options available to extend existing controls and classes. If you need a specific type of object to work with, there is a good chance the framework already has at least one commonly used form. We already mentioned API and .NET Framework options for Windows applications and commonly used controls available. Take some time to check out framework options before considering custom or third party choices.

The following table will help make you aware of some of these commonly used features. Included with this reference is a namespace reference, when applicable. Also, you will see references to .exe tools. These tools maybe part of the framework toolset already bundled into the installer, but some will require a separate download. Part of the features listed here are .NET concepts that have also been provided that will be worth looking into, if you plan to build enterprise applications.

TIP: Search out the .exe names provided to learn more.

Framework Tools

Table 6 Framework Options	
Tool	*Filename*
.NET Framework Command-Line Debugger	MDbg.exe
.NET Framework Configuration	Mscorcfg.msc
.NET Security Annotator	SecAnnotate.exe

**Table 6 Continued
Framework Options**

Tool	Filename
.NET Services Installation	Regsvcs.exe
ASP.NET Compilation	Aspnet_compiler .exe
Assembly Binding Log Viewer	Fuslogvw.exe
Assembly Cache Viewer	Shfusion.dll
Assembly Linker	Al.exe
Assembly Registration	Regasm.exe
Certificate Creation	Makecert.exe
Certificate Manager	Certmgr.exe
CLR Version	Clrver.exe
Code Access Security Policy	Caspol.exe
Code Generation	SqlMetal.exe
CorFlags Conversion	CorFlags.exe
Global Assembly Cache	Gacutil.exe
Installer	Installutil.exe
Isolated Storage	Storeadm.exe
License Compiler	Lc.exe
Management Strongly Typed Class Gen	Mgmtclassgen.exe
Manifest Generation and Editing Tool	Mage.exe
Manifest Generation and Editing	MageUI.exe
MSIL Assembler	Ilasm.exe
MSIL Disassembler	Ildasm.exe
Native Image Generator	Ngen.exe

Table 6 Continued Framework Options	
Tool	*Filename*
PEVerify	Peverify.exe
Resource File Generator	Resgen.exe
Set Registry	Setreg.exe
Sign Tool	SignTool.exe
Software Publisher Certificate Test	Cert2spc.exe
SOS Debugging Extension	SOS.dll
Strong Name	Sn.exe
Type Library Exporter	Tlbexp.exe
Type Library Importer	Tlbimp.exe
Windows Forms ActiveX Control Importer	Aximp.exe
Windows Forms Class Viewer	Wincv.exe
Windows Forms Resource Editor	Winres.exe
XML Schema Definition	Xsd.exe
XML Serializer Generator	Sgen.exe
XML	System.Xml

Source: *Microsoft.com,* ".NET Framework 4.5,".*NET Framework Tools* http://msdn.microsoft.com/en-us/library/d9kh6s92.aspx, 05/24/2012.

XML (*System.Xml*)

You can work with eXtensible Markup Language (XML) using Visual C# because the environment provides native support for it. eXtensible Markup Language helps you transport and store data. Keep in mind, though, .NET uses Microsoft XML Core Services (MSXML).

XML tools:

- XML Schema Definition Tool (*Xsd.exe*)
- XML Serializer Generator Tool (*Sgen.exe*)

Objects available include:

- Get to DOM (*System.Xml.XmlDocument*)
- XslTransform (*System.Xml.Xsl*)
- XPathNavigator (*System.Xml.XPath*)
- XPathDocument (*System.Xml.XPath*)
- XmlReader (*System.Xml*)
- XmlTextReader (*System.Xml*)
- XmlWriter (*System.Xml*)
- XmlTextWriter (*System.Xml*)
- XmlNode (*System.Xml*)

SOAP

The framework has Simple Object Access Protocol (SOAP) support. This simple and extensible XML based protocol is intended to help applications exchange data using HTTP. It is great for working around firewalls, because most firewalls allow using HTTP. Open the Microsoft SOAP Toolkit for enhanced data exchange.

REST

The framework has native Representational State Transfer (REST) support through the Windows Communication Foundation (WCF). This is a simpler form of web service. Many find it an advantageous replacement to SOAP and WSDL based design when you do not need all the good stuff these other interfaces can provide.

TIP: Search for ".NET REST WCF" and "RESTful" to learn more about this subject.

Membership

You may be familiar with the concept of Membership. This a framework option for managing users and groups. If used properly, it is especially helpful for managing Intranet applications.

TIP: Search for .NET Managing Users by Membership to learn more about this subject.

Directory Services (*System.DirectoryServices*):

If you are not familiar with Active Directory, it is used in network administration and security management. The *DirectoryServices* namespace is your vehicle for Active Directory and LDAP management. Other related sub classes include:

- *System.DirectoryServices.ActiveDirectory*
- *System.DirectoryServices.Protocols*

Do Asychronous (*System.ComponentModel*)

Asynchronization is a creative way to improve performance, to control processing, and to modularize computations. The *ComponentModel* provides the hook you need to implement these simultaneous computing requests.

Serialize (*System.Runtime.Serialization*)

When you need to serialize your data, *Serialization* is there to get the job done. You will find this object easy to work with and easy to implement.

XAML (*System.Xaml*)

This is part of the Windows Presentation Foundation (WPF). Objects included here are:

- XamlReader (*System.Xaml*)
- XamlObjectWriter (*System.Xaml*)
- XamlObjectReader (*System.Xaml*)
- XamlXmlWriter (*System.Xaml*)

TIP: Search for XAML namespace to learn more about how to work with this technology.

Thread Management (*System.Threading*)

Microsoft Windows computing operations are processed via the thread. You can manage this operating system process and its available objects include:

- ThreadPool (*System.Threading.ThreadPool*)

- BackgroundWorker
  (*System.ComponentModel.BackgroundWorker*)
- Monitor (*System.Threading.Monitor*)

Interop (*System.Runtime.InteropServices*)

There are a couple of highly used classifications within Interop: unmanaged code, and COM. As you get into COM remember to check out *Regasm.exe*, the Assembly Registration Tool.

Caching (*System.Runtime.Caching*)

Hook into the Microsoft Windows data store to improve your data management. When you need a good place to temporarily save your frequently used information, then look at what this component can do for you.

Unit Testing

Depending on the tools and the version you purchase, there could be built-in, unit testing features. Before you buy, review which features are included. If the version you are considering doesn't have native unit testing support, you may need to consider a version upgrade or a third party product. A later chapter covers more details on testing.

Going Custom

Web and Windows applications offer many customization options and the framework allows for implementing custom or extended classes. It also provides great hooks for working with custom controls.

Other ASP.NET Tools

- Browser Reg. Tool (*Aspnet_regbrowsers.exe*)
- Compilation Tool (*Aspnet_compiler.exe*)
- IIS Registration Tool (*Aspnet_regiis.exe*)
- Merge Tool (*Aspnet_merge.exe*)
- SQL Server Reg. Tool (*Aspnet_regsql.exe*)

Reminder: As discussed earlier, Web Services and ADO.NET are options for other forms of data management native to the Microsoft .NET Framework. Using Microsoft SQL Server support with the tools to get at Office data via Excel and Access, you can work with many of these common data systems.

Chapter 10

Search Engines are Your Friend

If you have not figured it out yet, Nerds are big fans of search engines, which are great avenues for finding information quickly provided you are Internet linked. Most importantly, they can connect you to people outside your normal circle of influence. As you associate with fellow Nerds, you can find answers to complicated problems, which might otherwise go unsolved. We are talking about *coding* problems here. (Fellow Nerds may or may not be able to help you with your personal issues.)

This book is written from the perspective of not being about what you know, but about knowing what to look for, and where to find it. All the code references are hints for you to venture away from your reference materials and explore the Internet for answers and deeper understanding. They are also here to inspire you to "network" with fellow Nerds, if you will forgive the pun.

Here are some keys to success. First, you should trap your coding error (or properly describe your problem), then copy and paste it into an Internet search to find how others resolve the same issue. One way to extract and error is to use *try{ } catch{ }*. Another interactive method is to trap Visual Studio error information via Breakpoint and Quick Watch.

TIP: Less interactive options to trap and report errors includes use of message boxes, popup windows, event logs, and custom handling. You can fix about 98.5 percent of the coding issues you run into by combining an error trapping technique with an Internet search. (The 98.5 percentage is a rational guestimate based on the alignment of the stars and the current rotation of the earth and the sun.) If your a non-Nerd or a wannabe Nerd we are trying to queue you in on the important stuff, however, you titled Nerds should already know this.

To address the remaining 1.5 percent of issues, what aren't addressed in a blog online, start your own blog post. Preferably, choose one of the more popular Visual C# sites. The higher the member count, the better your chances are for someone to respond.

Expect to resolve the majority of your coding errors via online research. As a manager, this a good reason all Nerd's should have full access to the Internet. This would include the *computer* and *blog* categories. They really do need it to handle the job. Also, if the Internet crashed in your office, and the Nerds all look at you with a perplexed look and chant "we need Internet," then assume they are telling you the truth and get them busy fixing it.

Moving forward...the primary reason Internet research is vital is because reference materials don't do a great job of explaining the necessary "how-to" details of using and implementing code in the real world. Blog posts are great because they tend to focus discussion on that crucial how-to part, whereas printed materials miss the mark.

If and when the Internet fails you, consider the following options:

- Search under a different concept.
- Try a different form of implementation.
- Talk with the vendor.
- Consider a third party consultant.
- Engage other Nerds in a more personal way (not just a blog post).
- Get creative.

Before we move off this topic, there is one more search related idea to address. This falls under the category of style guides, more specifically - how to name conventions. Please do a little research on the best practices of naming conventions, and put a little thought, intuition, and creativity into how things are named.

Other Nerds reading your code will appreciate the use of *intuitive* naming. If you are not sure what intuitive is, then search it out on the Internet. Keep reading until you get it. I stress this because of he who named his form object a "widget." Nerds used to be able to talk to non-Nerds and use this word "widget" to describe a hypothetical object and get the point across. Then a Nerd called a real object a widget and since then, non-Nerds hear "widget" in the context of something that is real. This confused hypothetical concepts with real ones. Many non-Nerds can't process this kind of abstract complexity. Now what do we call our theoretical objects? Understand Nerd power, and use it wisely.

Chapter 11

Testing Time

There are many ways to test code. This section will cover unit testing. Many of you senior Nerds have probably experienced the joy of testing. If you are familiar with unit testing I'm sure you know that this is a particularly special time in the software development life cycle. It is the part of the cycle designed to destroy your pride, humiliate you in front of other Nerds, and ultimately instill character development.

Endure unit testing because the end product of a good test run is hardened, quality, and slightly less buggy code. It can be a grueling process, especially if you are working on a project in which your code changes have to pass the inspection of 100,000 tests.

If you are unfamiliar with unit testing, its goal is to evaluate individual sections of source code. The basic goal is to provide some evaluation of code to see if it behaves as intended. It is also helpful to throw some unexpected situations at the process to see if the end result is the desired event.

Some planning is required to build a comprehensive set of tests. Do not hesitate to put in some good planning time into the test building part of the development life cycle.

There are a several different ways to evaluate a successful unit test:

- *Assert* evaluation
- No exception
- Caught exception using *ExpectedExceptionAttribute*

It is possible to combine both *Assert* and the *ExpectedExceptionAttribute* into one unit test. In this combination option either event can cause a failure, so build your unit test carefully.

TIP: Get to Unit Tests in your tool via the Test menu. To run an existing test, click Test > Windows, and then select Test View. Once the Test View window is displayed, right-click Test, and click Run Selection.

Assert (Microsoft.VisualStudio.TestTools.UnitTesting)

Before we jump into working with *Assert*, I am often asked about how users should go with unit testing. My recommendation is to build tests so that every section of your processing code is covered under a unit test of some type. I prefer building quality products and favor a very conservative approach, though that may not be appropriate for every situation. Consider the resources you can commit to testing and use your judgement. I like putting time into unit tests because they are cheaply reused throughout the application life cycle, giving you significant mileage for your investment.

Explore things you can do with *Assert*:

- *Assert.AreEqual*
- *Assert.AreNotEqual*
- *Assert.AreNotSame*
- *Assert.AreSame*
- *Assert.Fails*
- *Assert.GetType*
- *Assert.IsFalse*
- *Assert.IsInstanceOfType*
- *Assert.IsNotInstanceOfType*
- *Assert.IsNotNull*

- *Assert.IsNull*
- *Assert.IsTrue*
- *Assert.ReferenceEquals*
- *Assert.ToString*

Assert in action:

```
...
bool MyCustomer = true;
bool YourCustomer = false;
// This test will fail
Assert.IsTrue(MyCustomer,YourCustomer);
...
/* Catch the error if you want to do
 * further evaluation.
 ***********************************/
try { Assert.IsTrue(MyCustomer,YourCustomer); } catch { ... }
...
```

ExpectedExceptionAttribute (Microsoft.VisualStudio.TestTools.UnitTesting)

This thread safe attribute will identify whether an expected exception emerged during your unit test. The test will pass if the exception thrown matches the defined expected exception type.

TIP: The *ExpectedExceptionAttribute* class cannot be inherited.

For example:

```
[TestMethod()]
[ExpectedException(
    typeof(NullReferenceException))]
public void MyTest() {
  // cause exception here ...
}
```

Build a Unit Test

Once you feel familiar with unit testing, you can put our shared knowledge together with some new information to build a working test. The next example will link our test class into the unit testing framework.

Below is a practical example of a unit test:

```
using MyRealCode;
using Microsoft.VisualStudio.TestTools.UnitTesting;
namespace OurTestProject {
[TestClass()]
public class RealCodeTest {
private TestContext testContextInstance;
[ClassInitialize()]
public static void ClassInit(TestContext context) {...}
[TestInitialize()]
public void Initialize() {...}
public TestContext TestContext {
get {...}
set {...}
[TestMethod()]
public void TestMethod1() {
//build accessor to real code to test
MyRealCode target = new MyRealCode();
...
try {
somevalue = target.RealCodeMethod(...);
Assert.IsNull(somevalue);} catch {...}}
[TestCleanup()]
public void Cleanup() {...}
[ClassCleanup()]
public static void ClassCleanup(){...}
}}
```

TIP: A quick way to get a shell unit test started is to open your C# class (.cs) file and right click anywhere in the open file. Then, choose the "Create Unit Tests" option and check the options you want to add to your unit test file. If you do not have a unit test repository, be sure to add a "Test Project" to your working Solution. To get to private methods in your code, you will need to use an *Accessor*. These unreachable code sections become accessible using the black box. Take a close look at the **target** variable in the example above.

Another vehicle to expose visibility of Internal and Friend methods is to make use of the *InternalsVisibleToAttribute*. This attribute allows you to specify other classes that are allowed to come out. Keep in mind that both assemblies must have the same type of signing, either unsigned or strong name.

To build a unit test without source code, you can create a Test Project. When the new project is open, access the Create Unit Test dialog and use the Add Assembly option. From there you will be able to construct your test around an existing assembly.

Testing is something you should consider essential to the whole life of your software, and it should be integrated into all stages of the software life cycle. Consider building tests into the requirements, design, coding, and maintenance stages. Inside each stage the test itself has a life cycle.

Tests have requirements, acceptance criteria, design, coding and executing, and maintenance. Testing should undergo regular review and auditing with acceptance and baseline benchmarks. Consider unit tests as one of the most important testing vehicles a Nerd has at their disposal to ensure software operates as it should.

TIP: This unit testing talk not nerdy enough for you? Search out how to integrate XML Serialization into your testing process. Use it to test those classes that can be serialized.

Troubleshooting

Once in a while things go really badly. When they do, I like to turn on logging for the Integrated Development Environment (IDE). Sometimes the IDE tools cannot handle the working tasks. If the IDE has problems, you may catch errors if you have logging turned on. Do this by opening the *devenv.exe.config* file in your IDE directory, then adding this:

```
<configuration>
...
<system.diagnostics>
    <trace autoflush="true" indentsize="4">
      <listeners>
        <add name="myListener"
type="System.Diagnostics.TextWriterTraceListener, System
version=1.0.3300.0,Culture=neutral, PublicKeyToken=b77a5c561934e089"
initializeData="c:\myListener.log" />
        <remove name="Default" />
      </listeners>
    </trace>
  </system.diagnostics>
</configuration>
```

TIP: Older IDE environments may have a different configuration so search for *devenv.exe.config trace* or *KB 950609* to find your settings.

Chapter 12

Help Wanted. Join a Nerd Community

Whether you want to be a Nerd or already are one who wants to become an extraordinary Nerd, then consider "networking" with other Nerds–pun intended. By associating with other Nerds you may achieve an increased synergy in your Nerd development or may help some other Nerd by your virtual or physical presence (and both of these count).

Relationship development can be time consuming, challenging, and may require some type of direct interaction with other human beings. If you cannot handle that, or simply need to build up the tolerance for such actions, then consider blogs, chat, and some other online communication vehicles to make your connect. Regular participation with both questions and answers usually will lead to a tolerance of your existence in these places of gathering. Hey, you might even become valued!

Nerds associate in several ways. Some of them will associate for free if the topic of conversation is nerdy, or interesting enough. Others try to be elite by separation through capital. Which means you can connect freely if you don't mind joining "the group" for a small fee. The real smart ones will allow your communication with them if you don't mind paying them for their time. Buy them some good "java" (English translation: coffee) and they may throw you an extra bone once in a while.

Tapping Nerd power is another way to address that pesky 1.5 percent of issues you can't easily search out an answer for. Ways to connect include:

a) Subscribe to and utilize one or more of these things called a Public User Group. We are talking technology platform here. It is the old school way in which Nerds originally communicated with one another during the dawn of the Internet age. Funny thing is that it is still in frequent use today in a seemingly preferred form of communication with one senior Nerd to another. Find one and sign up, then transmit your question.

b) Join a local User Community. This would be a regular meeting of local Nerds in your area. These groups are usually formed around mainstream technology platforms, so try an Internet search for .Net Group YourCity, YourState to see what could be near you. Also check your local paper for advertisements or see what your local IT University may offer. If you look hard enough, you will find the Nerds all meeting together somewhere eating pizza and talking about nerdy stuff, of course.

If you're a Nerd that really wants to go to the next level, and elite is just not good enough, then consider joining a Nerd association.

Chapter 13

In Case You Already Know Java

Here is a Java to C# syntax conversion table, which highlights frequently used differences between Java and Visual C#.

Table 7 Java vs. Visual C#	
Java	*Visual C#*
.finalize()	[not needed]
boolean	bool
class ME extends OTHER	class ME: OTHER
FileInputStream	BinaryReader
FileOutputStream	BinaryWriter
FileReader	StreamReader (System.IO)
FileWriter	StreamWriter (System. IO)
final	const
for (int y :myList)	foreach (int y in myList)
import	using
instanceof	is
interface ME extends OTHER	interface ME : OTHER
myString.compareTo()	myString.ComopareTo()
myString.equals()	myString.Equals()
native	extem
Object	object

Table 7 Continued Java vs. Visual C#	
Java	*Visual C#*
package	namespace
protected	internal
public int getME()	public int ME { get {} set {}}
public void setME ()	public int ME { get {} set {}}
String	string
String myList[] = new String[2]	string[] myList = new string[2]
StringBuffer	StringBuilder (System.Text)
super	base
switch () [limited types]	switch () [unlimited types]
synchronized	lock
System.in.read()	Console.Read()
System.out.printin()	Console.Writeline()

Source:*Daniel Diaz.* 2012

What a huge list! Don't be surprised at how far your knowledge of Java syntax will get you in the Visual C# world. Read something different if you want to get deeper than this.

A few years ago, there was an accusation made in the news several years ago that Visual C# was a copy of Java. The point being both languages do have similarities, thus the short list.

TIP: To see the library/namespace differences between Java and C# search out Java vs C# Library.

Good news for those pure open source developers. The tools you have been using for Java usually have a version for Visual C# coding. Check with the vendor before you give up a perfectly good software license.

TIP: Search for Java C# Conversion or Java to C# Converter to find tools that will automatically convert your existing Java code to C#.

Chapter 14

Hiring a Visual C# Nerd Interview Q's

This chapter is to provide guidance to those looking to hire Visual C# Nerds, and to lend some help to those pretending to be coders. We need to allow some freedom to these wannabes, so that they can blend in with the rest, talking like they know the game. Increasing your knowledge is important for asking the right questions as you grill prospective Nerds. The goal with this suggested interview process is not see if you can trick a Nerd with twisted questions, but rather to really find those who understand this stuff.

Web Specific

- Q: What is *ViewState*?
  A: *ViewState*, when enabled, will manage postback data. Form fields using this feature will be repopulated after a postback.

- Q: What is a major drawback to using *ViewState*?
  A: It takes up extra resources when the application does not require it. No *ViewState* will result in improved application performance.

- Q: Describe the client / server model.
  A: Can they articulate anything that resembles the information we discussed in the *You are About to Write Code* chapter?

- Q: What two classes were provided to allow for application debugging?
  A: The *Debug* and *Trace* class.

- Q: What is the simplest way to enable these two classes (*Debug*, *Trace*) in your application?
  A: Syntax for *web.config* file includes
      <system.web>

```
<compilation debug="true"...
<trace enabled="trace"...
```

- Q: What is a Theme and how is it used in ASP.NET?
  A: Themes are a collection of styling elements such as skins, cascading style sheets, images, and other stuff. They can be turned ON at the Application level in the *web.config* file, and at the Page level by setting *Theme* to an available theme name. Form elements can also access themes through *ThemeableAttribute.*

- Q: How can you deploy an ASP.NET web application to an external client?
  A: Through the use of an Installer project.

- Q: How do you encrypt connection string info inside a *web.config* file?
  A: Through the use of *aspnet_regiis.exe* with the *-pe* and *-app* options.

- Q: How would you programmatically access IIS System settings?
  A: Through the use of one of the following vehicles: *ADSI*, *WMI*, or *COM*.

- Q: What is an application pool?
  A: An application group that contains its own worker processes. The pool is a boundary that separates each work group from one another.

For Web Or Windows Applications

- Q: What is a *Property*?
  A: Through the use of *get{} set {}* syntax there are a combination of methods and variables that extend a class to provide the ability to read, write, and compute values.

- Q: What is the basic syntax for trapping errors?
  A: The most basic syntax form is *try{} catch{}* while a slightly more complex form includes *try{} catch{} finally{}*.

- Q: How is DLL Hell solved in .NET?
  A: Assembly versioning allows the application to specify the library and the version of the assembly needed.

- Q: How do you make a DLL in C#?
  A: Use the */target:library* compiler option or build a *Class Library*.

- Q: How is a method overloaded?
  A: Via changed parameter options. There can be different data types, numbers of parameters, or orders of parameters.

- Q: What is a constructor?
  A: A method in the class which has the same name as the class and is initialized whenever an instance of the class is created.

- Q: Name the four main access modifiers in C#.
  A: *public, private, internal, protected*

- Q: What's an abstract class?
  A: A class that must be inherited and have its methods overridden. This class type cannot be instantiated and is the inheriting class pattern.

- Q: Is there a call to force garbage collection?
  A: Yes, *System.GC.Collect()*.

- Q: Do you know what REGEX is?
  A: Regular expressions.

- Q: Is there a namespace available for REGEX support?
  A: *System.Text.RegularExpressions*

- Q: What is the top .NET class that everything is derived from?
  A: *System.Object*

- Q: Method name to clear an object from memory?
  A: The method is *Dispose*.

- Q: What is Reflection, and its purpose?
  A: It is a collection of classes that allow access to assembly metadata through namespace *System.Reflection*.

These questions are not tricky, but they will show you if the person sitting in front of you has any previous Visual C# coding experience. These are designed to represent frequently used concepts. However, you may have someone in front of you who has read this book, so mix it up a little. Try to keep the Nerds on their toes. There is nothing more destructive to society than a completely lazy Nerd.

TIP: You guessed it - search for C# Interview Questions to find more so you can mix it up.

Chapter 15

Nerd Compensation

Maybe you are a wannabe Nerd. Perhaps you believe the title of Nerd, will improve your financial or work status. Perhaps, you are already of Nerd status and think learning this Visual C# will allow you to achieve the position you have always wanted. We will dig into the data to see if more info about the job may help or hurt your consideration.

The average code Nerd's median salary is between $80,000 to $90,000 USD. If you are thinking that is a lot of money, it is important to remember that this is for the Nerd with more than five years experience and a bachelor's degree. If you do not have this type of Nerd experience, you are going to need to get started in order to achieve this type of salary.

Are you lacking a college degree? You will **really** need to work on that. The corporate world enjoys hiring and working with accredited Nerds. More and more Nerds are hitting the workforce, giving businesses the luxury of filtering out who they hire based on education and experience. The certification stuff alone does not sit well with hiring managers. The point being that it does not have the same effect on hiring managers and human resources as holding a college degree does.

It's time for a reality-check. Remember Nerd work is usually *hard*. It can be very stressful. Most code Nerds spend all day chained to a desk, which can have a detrimental effect on your health. Code Nerds use a keyboard and mouse and stare a computer screen all day long, which can also be hazardous to your health. If you stare at a computer screen all day every day for ten years, glasses [if you do not already have time] may be in your future. Have you heard of carpal tunnel syndrome? This syndrome can occur when you use a keyboard for an extended period of time without doing something to counteract the muscle motion.

Let's move on to the topic of mental anguish. Our first look is to the vendor that created .NET. In order to maintain this platform and improve it, the creators are constantly making changes. They do this by fixing the broken, making improvements, and adding new things.

Updates are usually released on a frequent basis. Their releases usually contain information you need to become familiar with in order to best complete your job. This is the first entry into the problem - regular learning dictated by the vendor at the vendors pace. We cannot fault the vendor because they are just trying to remain competitive by keeping pace with all this other stuff that is happing in the tech world.

The point is you have to continually learn or your competitiveness in the market place is likely to dwindle. This learning encompasses a absurd volume of information which is constantly changing. Continual learning at a fast pace can be very difficult even despite holding Nerd status. It is a real commitment and should not be taken lightly.

We had a non-Nerd read through this book prior to publication. Individual was under close observation as an experiment. The intent was to monitor emotional and physical reactions to this material. Prior to exposure they had a general interest in learning computer stuff. After the read, something changed, and a strong desire to keep their interest as an interest developed to the level where it became something not worth exploring. The indicated reason for the change was because the volume of information regardless of how much or how little that is presented in this work was just too much to take in. "Who really wants to bother learning all this stuff only to know that it may all be irrelevant knowledge next year." Perhaps, this is why Nerds get paid well. The industry is risky business requiring knowledge, creativity, and potential to endure physical and mental harm.

To prove our point that Nerd work is stressful we need a volunteer to pick on. You know about how volunteer service works in the military right? I want to thank writer Victoria Brienza for volunteering. In her article titled [3]*10 Least Stressful Jobs of 2011* we find that apparently (speculating here) she may not have asked very many programmers how stressful their job is. Perhaps she was paid to make programming look more glamorous than it really is? I would like to review her research data and random sampling methodology. Check it out and read the article comments, ROFL.

The picture of "the programmer" says it all! It reminded me of a guy I worked with. When surveyed about his morning food consumption in a meeting, he admitted he had consumed a Diet Coke for breakfast. This is when all others surveyed had eaten more normal stuff like cereal, eggs, toast, bagels, etc. The guy with the Diet Coke also had the *best* coding project out of the group. Like my Diet Coke friend who realized that his problem was not something that code alone could fix. Nerds eventually find out that coding is really about working on business problems. These problems are usually very complicated, and to make it worse, pure technical know-how holds little value in working out a good resolution. This can be very stressful when you are supposed to be the fix deliverer with a magic bullet in pocket. Coding is done to meet business needs.

For some good news, [4]computer software engineering is expected to be among the fastest-growing occupations through 2016, according to the Department of Labor. It is interesting that this statistic is valid until 2016, like it is a suggested transition year for this career path, business sector, or perhaps the economy as a whole.

Programming has been considered many times over as a top 10 recession-proof job.

Job security has been arguably the big driver of anyone in a technology career. Businesses constantly face difficult challenges. They often turn to technology to deal with their problems. Being part of the problem resolution is a good way to become a needed member of an organization. As long as the trend to be technology focused continues, the need for a code Nerd will be around with no guarantee as to what that Nerd's class status really should be.

Chapter 16

Code Reference Charts

Here are a couple of help guides that are good for quick reference.
- Code examples
- Operators

Basic Form Quick Reference
if (eval) { }
if (eval) { } else { }
if (eval) { } else { } if (eval) { } else { }
<eval> ? <if true> : <if false>
switch (eval) {case evalA: break; case evalB: break; default: break;}
for (initialize; condition; operation) { }
foreach (Type Item in List) { }
while (eval) { }
do { } while (eval)
break; [exit loop]
continue; [continue to next loop]
Type[,] vArray = new Type [length]
Type [,] vMultiDimArray=new Type [length]
Type [] [] vJaggedAry = new Type [length] [ ]   vJaggedAry [0] = new Type [new length]
void ReturnNothing () { }
Type ReturnSomething () {return val}
enum Enum {val1, val2}
struct Struct {item1; item2;}

Basic Form Quick Reference Continued
class Class {}
Type Property { get { } set{ } }
namespace Namespace{ class Class { } }
class ChildInherits : Parent { }

Source: *Daniel Diaz*. 2012.

Operator	Purpose	Priority
.	member access	left-to-right
()	method call	
[]	element access	
++	postfix increment	
--	postfix decrement	
new	object creation	
typeof	get the type of object	
sizeof	get the size in bytes of a type	
+	unary plus	
-	unary minus	right-to-left
!	logical negation	
~	bitwise not operator	
++	prefix increment	right-to-left
--	prefix decrement	
(type)	casting	
*	multiplication	left-to-right

/	division	
%	modulus/remainder	
+	addition	
-	subtraction	
<<	left shift operator	
>>	right shift operator	
<	less than	
>	greater than	
<=	less than or equal to	
is	type comparison	
as	type conversion	
!=	is not equal to	
==	is equal to	
&	bitwise AND	
^	bitwise XOR	
|	bitwise OR	left-to-right
&&	logical AND	
||	logical OR	
=	assignment	right-to-left
*=	multiplication assignment	right-to-left
/=	division assignment	
%=	modulus assignment	
+=	addition assignment	
-=	subtraction assignment	
<<=	left shift assignment	

&=	bitwise AND assignment	
^=	bitwise XOR assignment	
|=	bitwise OR assignment	

Endnotes

1 *Microsoft Developer Network.* http://msdn.microsoft.com/en-us/library/dct97kc3.aspx
taken on 5/14/2012.

2 *Blog by Herman Gupta.* http://hemantg.blogspot.com/2010/12/aspnet-event-sequence.html taken on 5/14/2012.

3 *CareerCast.* http://www.careercast.com/content/10-least-stressful-jobs-2011-4-computer-programmer taken on 5/14/2012.

4 *Forbes.com.* http://www.forbes.com/2008/07/18/recession-proff-jobs-lead-careers-cx_tw_0718recessionproof_slide_3.html? taken on 5/14/2012.

About the Author

Globally recognized leader and executive, Daniel Diaz III©™ is an expert in the high tech industry operating under his specialty of B2B software. He is also an expert technologist and a recognized Nerd. He is officially Nerd Guide's first Nerd Certified™ candidate. Some will recognize him as "TheTechExpert"™. Now he is an author for and founder of Nerd Guide™. Mr. Diaz holds over seventeen years of workforce experience.

His academic credentials include an undergraduate degree in Computer Science and a master's degree in Management. Add to that his industry experience which includes the aerospace, defense, software, finance, and high tech manufacturing sectors. His technology output is influenced by his holistic and analytical view of business, and his conservative investment strategies for maintaining positive cash flow. He is a real finance Nerd.

Mr. Diaz's technical competencies include proficiency with commonly used object oriented program languages such as C#, VB.Net and Java; and expertise with frequently used database systems like MySQL, Oracle RDBMS, and Microsoft SQL Server.

Throughout his career, Mr. Diaz has worked with public, private, non-profit, education, government, and military entities. He holds working knowledge of compliance and certification processes including banking regulation, CMMI, HIPPA, ITIL, and SOX. A summary of his talents includes:

- Executive Leader

- Architect

- Strategist

- Visionary

- Investor

- Fellow Nerd

As a part-time initiative, Mr. Diaz enjoys perpetuating technical education through the Nerd Guide project.

"When normal is not working for you, engage someone who is extraordinary."

About the Publisher

Nerd Guide is the premier publication and education source, providing technical knowledge, concepts, training, and documentation for Nerds. We hope to become a strong non-profit educational source for high tech learning. We look forward to providing materials for all ages.

We are looking for authors (aka Nerds), publishers, translators, bookstores, and other distribution outlets to help us perpetuate these great educational opportunities.

Is high tech education for all your passion? We except donations:

Nerd Guide
P.O. Box 15559
Rochester, New York

Yes, we are not a non-profit yet but that just means you cannot claim your donation as tax deductible. Be patient with us, we will get there, after all we just launched in 2012!

About Nerd Certification

Nerd Guide is continuing to develop its Nerd Certified™ program. Through this initiative, Nerd Guide hopes to provide a platform for Nerds to be certified and credentialed as an official Nerd. In addition to certifying Nerds, we hope to certify products as Nerd worthy through our developing Nerd network. We hope that as this network progresses you who are worthy of being called a Nerd will be willing to participate.

More information about this program will be provided on the website, found at nerdguide.org.

Nerd Guide Branding

To help boost our branding efforts Nerd Guide would like to connect with vendors who can help sell branded Nerd Certified™ and Nerd Guide™ merchandise. Visit our website to see products that are currently available.

Nerd Guide designed this fun, lighthearted program to grow revenues in order to support continued and expanded tech education for all ages.